Understanding the Elements of the Periodic Table™

THE NOBLE GASES

Helium, Neon, Argon, Krypton, Xenon, Radon

Adam Furgang

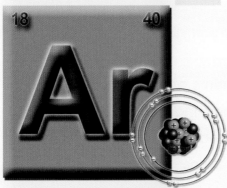

rosen publishing's
rosen central®

New York

*For my father, who introduced me to science and
let me peer into a microscope before I could even read*

Published in 2010 by The Rosen Publishing Group, Inc.
29 East 21st Street, New York, NY 10010

Copyright © 2010 by The Rosen Publishing Group, Inc.

First Edition

Library of Congress Cataloging-in-Publication Data

Furgang, Adam.
The noble gases: helium, neon, argon, krypton, xenon, radon / Adam
Furgang.—1st ed.
   p. cm.—(Understanding the elements of the periodic table)
Includes bibliographical references and index.
ISBN 978-1-4358-3558-0 (library binding)
1. Gases, Rare—Popular works. 2. Atoms—Popular works. 3. Periodic
law—Popular works. I. Title.
QD162.F87 2010
546'.75—dc22

                                                          2009014409

Manufactured in Malaysia
CPSIA Compliance Information: Batch #TW10YA: For Further Information contact Rosen Publishing, New York,
New York at 1-800-237-9932

On the cover: The six noble gases, as they appear on the periodic table.
The atomic structure of each element is shown.

Contents

Introduction

Have you ever looked around you and wondered where every-thing comes from? What is all the matter that you see in your environment made of? Every substance and material around you—every living and nonliving thing that you see every day—is made up of chemical elements.

Chemical elements are the simplest form that matter can be broken down into. When you talk about elements, you are really talking about atoms and how they are arranged. An atom is the smallest form of matter that exists. It cannot be broken down any further by chemical reactions. There are more than ninety elements found naturally on Earth. In addition to these elements, there are more than twenty that are made in laboratories. Every other substance on Earth is made of a combination of these elements.

An element in its pure form will only be made up of atoms from that element. For example, gold (Au) will only have gold atoms in it. Mercury (Hg) will only have mercury atoms in it. Some elements, such as silver (Ag) and iron (Fe), are commonly known and used in everyday life. Others elements, such as francium (Fr) or berkelium (Bk), are only known about, used, and understood by scientists. Some elements are commonly found all over Earth's surface, while others are rare and found only in small amounts in nature.

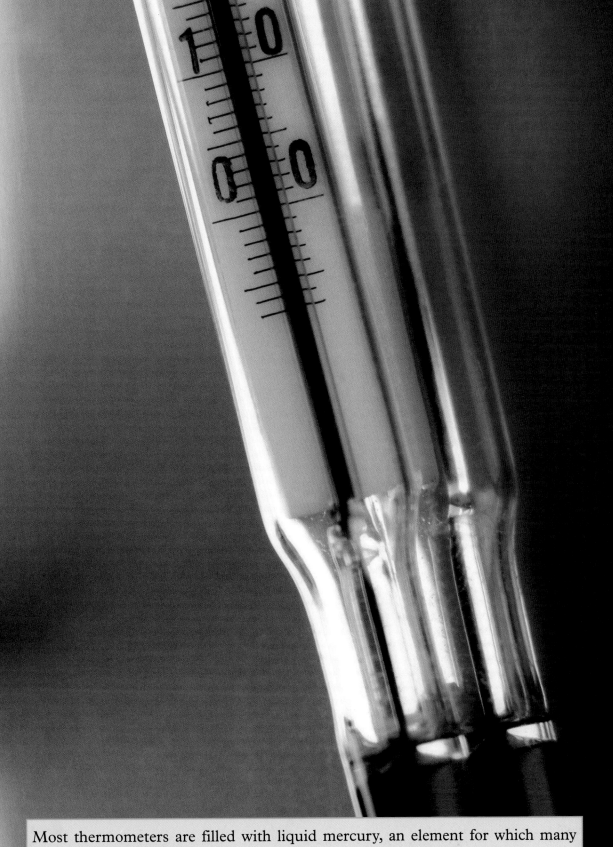

Most thermometers are filled with liquid mercury, an element for which many practical uses have been found.

How do scientists group elements? First, they look at the atoms that make up an element. Each atom has its own structure—an arrangement of parts that makes each element unique. The arrangement of the parts of an atom in any single element gives that element its own special properties and characteristics.

Among all the many elements, there are six elements that are gases and that are grouped together by scientists. These are the noble gases. The noble gases are a unique grouping of elements found naturally on Earth. These elements are helium (He), neon (Ne), argon (Ar), krypton (Kr), xenon (Xe), and radon (Rn).

The noble gases are mixed in with the air that we breathe. Some of the gases are more common than others. For example, helium is one of the most common elements in the universe, but it is found only in limited amounts in Earth's atmosphere. Throughout this book, you will learn about what makes the noble gases unique. You will also learn about how their atoms are arranged and how they are used in industry, our daily lives, and scientific research.

Chapter One
Elements and Their Parts

To understand how one element differs from another, you must first understand what makes up an atom. By performing many experiments, scientists have been able to discover different parts of an atom and uncover some of the mysteries of what makes each element unique.

Protons, Neutrons, and Electrons

Atoms make up elements. They are very small, but they are not the smallest known particles. Atoms themselves are made up of even smaller particles. If you could go to the center, or core, of each atom, you would find its nucleus. The nucleus is made of tiny particles called protons and neutrons. Protons have a positive electrical charge, while neutrons have no electrical charge. Each element has a different number of protons at its core. This number of protons in an atom of an element is called its atomic number.

If you look at the outer shell of an atom, you will find tiny particles called electrons moving around inside the atom. The number of electrons in an atom matches the number of protons in that atom's nucleus, but the electrons have a negative electrical charge.

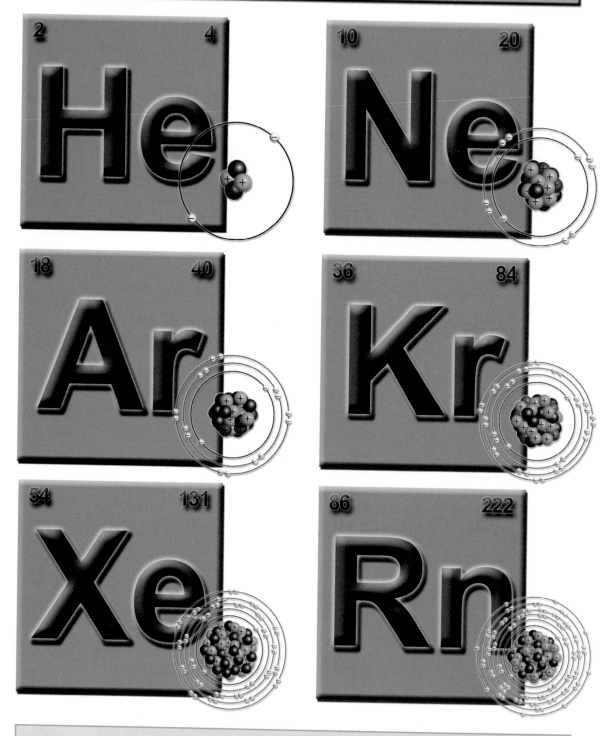

These are the symbols that represent the six noble gases. Each element shown here has an atomic number, an atomic weight, and a visual representation of its atom.

For example, the atomic number of hydrogen (H) is 1. That means a hydrogen atom has one proton and one electron. The atomic number of gold is 79. That means a gold atom has seventy-nine protons and seventy-nine electrons.

Each element has a different arrangement of electrons around the nucleus. The arrangement of electrons forms a kind of shell around the atom. The outer shell of an atom, where the electrons are, is about ten thousand times larger than the nucleus. What holds the parts of an atom together? An electromagnetic force holds an atom together and keeps it from flying apart. This is similar to the gravitational force that keeps objects around us firmly on the ground. The same force keeps the planets in a stable orbit, rather than spinning off on a wild trajectory. It is the atoms that give mass to every substance around us. The atoms of

Dmitry Mendeleyev

Dmitry Mendeleyev was born in Siberia, Russia, in 1834. He was the youngest in a family that had as many as fourteen children. While Mendeleyev was finishing high school, his father died. The family's glass factory burned down soon after. His mother took him to St. Petersburg, where he enrolled in the Main Pedagogical Institute. He studied science and graduated in 1856. Mendeleyev is best known for his work on the periodic table of elements. In 1869, he presented a paper before the Russian Chemical Society in which he introduced his table of the sixty-three known elements. Each element was listed according to its atomic mass. The basis of his work is still in use today in the modern periodic table. Mendeleyev died in 1907.

different elements have different masses. This is called the element's atomic mass. The atoms of lead (Pb) are much heavier than the atoms of aluminum (Al), and this is why a piece of lead is much heavier than a piece of aluminum of the same size.

Getting Organized

Because there are so many elements, each with its own properties, scientists began trying to organize them in some logical way.

Russian scientist Dmitry Mendeleyev is credited with creating the first periodic table that logically arranged the known chemical elements. First, he arranged similar elements together. Eventually, he organized the elements according to each one's atomic mass. The modern periodic table is still based on Mendeleyev's original work. Today, the periodic table is slightly different from Mendeleyev's because now the elements are listed in order of increasing atomic number. The number of protons that each element's atoms have corresponds to the element's atomic number on the periodic table.

The elements on the periodic table all have different properties that make them distinct, though some can be similar to each other and thus grouped together. All the similar types of elements on the periodic table are listed within columns, or groups, each with its own number. For example, the elements in group 1 share certain characteristics. They are different from the elements that are in group 5.

On the far right-hand side of the periodic table is a group of six gases listed in a single column: helium, neon, argon, krypton, xenon, and radon. These are the noble gases. The noble gases make up one of the most interesting groups of elements on the periodic table. They stand out, even among other gases. The noble gases are listed on the periodic table near the other gases, but they have their own column, group 0.

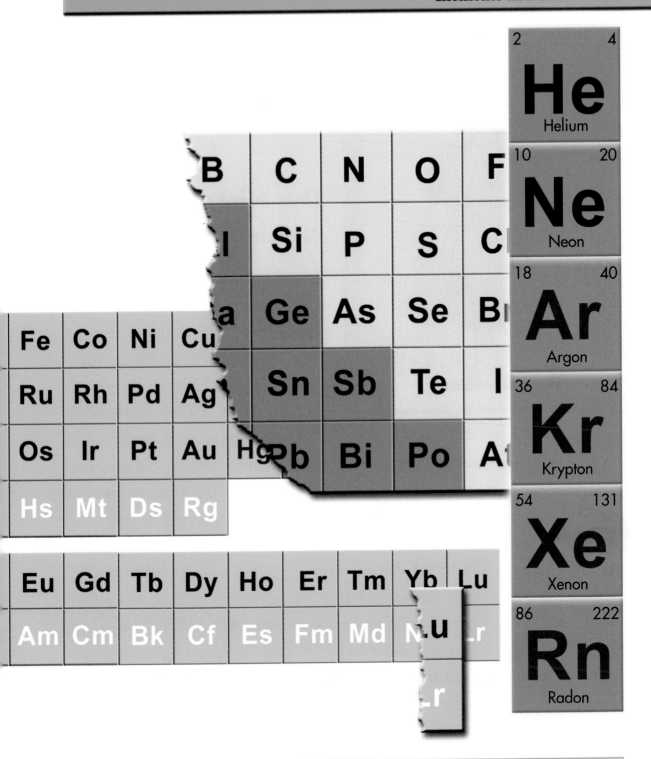

This is part of the periodic table of the elements. The six noble gases are located on the far right column of the table.

British scientist Henry Cavendish accidentally discovered the noble gas argon while conducting an experiment to create nitrous oxide.

The History of the Noble Elements

Scientists have not known about the noble gas elements for as long as they have known about some of the other elements on the periodic table. Back in 1785, a British scientist named Henry Cavendish was conducting an experiment in which he passed electricity through air from which carbon dioxide had been removed. He was hoping to create nitrous oxide, a compound of nitrogen (N) and oxygen (O). When all of the nitrogen and oxygen had reacted, an unexplained amount of gas remained that was neither nitrogen nor oxygen. Cavendish did not know it at the time, but this mystery gas was argon, one of the noble gases. However, it would be more than one hundred years before his discovery would be explained and classified.

In 1894, Scottish chemist Sir William Ramsay and English physicist Lord Rayleigh removed nitrogen from the air during an experiment. Like Cavendish, they, too, were left with a gas that they could not identify or explain. They concluded that they had discovered a new element. Ramsay called it "argon" from the Greek word *argos*, which means "lazy." He characterized the gas as lazy because it would not react with the other elements. He went on to discover some of the other noble gases: neon,

Today, scientists can use a gas spectrometer to detect which elements are present in a substance.

krypton, and xenon. Ramsay won the Nobel Prize in Chemistry in 1904 for his discovery of the noble gases, none of which combined chemically with the other elements on the periodic table.

The first noble gas ever to be identified was not even discovered on Earth but in the sun. In 1868, French astronomer Pierre Janssen and English scientist Joseph Lockyer discovered helium. One day, Janssen was observing a solar eclipse. He was using a spectroscope, an instrument that measures light from different portions of the electromagnetic spectrum. The light produced by different substances can help scientists determine what elements are present in these substances.

What Is a Gas?

A gas is one of the three states of matter. Matter can exist as a solid, liquid, or gas. Most gases have no particular shape. The groups of atoms that make up a gas are farther apart from each other than the groups of atoms that make up solids or liquids. A gas will spread out in its surroundings to evenly fill the space that it is contained within. For example, gas in a container—a small canister, a room, an underground cavern—will spread out to fill that container evenly. The air that we breathe is made up of many gases, including nitrogen and oxygen.

While observing the eclipse, Janssen saw light in the spectrum that was evidence of a new element. This new element was helium. Joseph Lockyer also observed and recorded this phenomenon at the very same time, and both men are credited with the discovery of helium. Lockyer named helium after the Greek word for the sun, *helios*. Years later, during his groundbreaking work with the noble gases, Sir William Ramsay isolated helium gas found here on Earth.

In 1900, German physicist Friedrich Ernst Dorn discovered radon while observing the radioactive decay of the element radium (Ra). Radon is produced when radium decays, although the two elements are completely different from each other. In 1910, Ramsay also isolated and described radon.

Chapter Two
Understanding the Noble Gases

To understand what makes the noble gases unique, it helps to think about the tiny atoms that make up each of these elements. The atoms of most elements have more than one shell of electrons around them. The outer shell of electrons is important because if it is not full the atom will combine or react more easily with other atoms around it to get a filled outer shell. This process forms chemical compounds.

A compound is a chemical combination of two or more elements that form a new substance. The atoms of all six noble gases have complete outer shells of electrons. This makes them very unreactive with other elements. They do not combine with other elements naturally or readily because their electrons do not react easily with electrons of other elements. Another term to describe the nonreactive nature of noble gases is "inert," which literally means idle, unmoving, or resistant.

It's unusual to find elements like the noble gases that do not react easily with other elements. Most substances around us are made from elements that have formed compounds. When elements combine chemically to form compounds, they lose their individual properties and form a new substance with its own properties.

For example, a very common and important compound found on Earth is water. Water is a compound made from combining the elements hydrogen and oxygen. The atomic symbol for hydrogen is H. The atomic

Water is a common compound on Earth. This computer model of a water molecule shows hydrogen in blue and oxygen in red.

symbol for oxygen is O. When atoms from different elements are combined, they form molecules. The symbol for a water molecule is H_2O. This means every water molecule contains two hydrogen atoms and one oxygen atom. A mixture of the elements hydrogen and oxygen is very explosive, but after they have reacted they form a stable and life-giving compound that is essential to the growth and survival of all life forms.

Originally, scientists thought that making compounds from noble gases was impossible due to their nonreactive nature. They described these elements as "keeping to themselves" and "amongst their own kind," like snobbish noblemen from the days of aristocracy. This is how they got their name: noble gases. Over the years, however, scientists who like a tough challenge have put these elements to the test and have managed to form compounds with some of them in laboratories. However, this is not done easily, and the resulting compounds are often not very stable.

The Noble Gases

All six noble gases have some similarities. They are all odorless, colorless, and tasteless gases. They do not combine easily to form compounds because of the complete outer shell of electrons that each element has. Unlike other gas elements, the noble gases will not even react with each other. All other gases on the periodic table will form a bond with one another at room temperature. The noble gases are so nonreactive that they cannot even form bonds with themselves.

How do the six noble gases differ from one another? Each has its own characteristics and practical uses. Read more about each of these elements below.

Helium

Symbol: He • Atomic number: 2

Helium is the second most abundant element in the universe, with hydrogen being the first. The entire mass of the universe is made up of mostly hydrogen and helium. Hydrogen makes up about 75 percent of the universe, while helium represents 23 percent. All other elements make up the remaining 2 percent of the mass of the entire universe! Despite being extremely common, helium is not easily found on Earth. A very tiny portion of Earth's atmosphere, only 0.00052 percent, is helium. Helium was first discovered in the sun because it is very abundant in the stars.

Helium is the first noble gas on the periodic table in group 0. This group falls along the far right-hand side of the periodic table. Helium is the

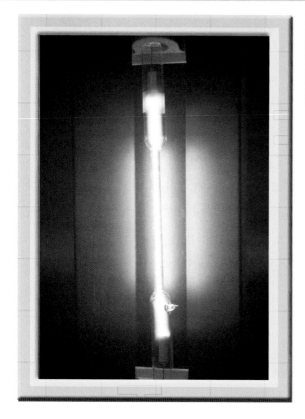

This pressurized helium emission tube shows the light and color that the gas emits when electricity is passed through it.

only element in the entire table that cannot be made solid by reducing its temperature and compressing it. When it is cooled to -452.09 degrees Fahrenheit (-268.94 degrees Celsius), it will become a liquid known as helium I. In theory, the coldest possible temperature is absolute zero. Helium liquid exists just a few degrees above absolute zero. If cooled even a little bit closer to absolute zero, helium liquid will become completely motionless and begin to expand. At this extremely cold temperature, helium is often referred to as a superfluid, or helium II. In this state, helium acts in unique ways compared to other liquids. Helium II liquid has zero viscosity. Viscosity is the measure of the internal friction that different liquids or fluids have. Because helium II has zero viscosity, it will continue swirling indefinitely once it has been stirred! The lack of friction prevents its motion from ever slowing or stopping.

Neon

Symbol: Ne • Atomic number: 10

Sitting right below helium and above argon on the periodic table is neon. Neon comprises about 0.13 percent of the universe, making it the fifth most common element. It represents 0.0018 percent of the air that we

The most common use for neon is inside bright, colorful light tubes called neon lights. Although they are called neon lights, neon is actually mixed with other gases to achieve different colors.

breathe. On Earth, pure neon is obtained by extracting it from the atmosphere because there are no solid substances that contain it. Neon gas is naturally colorless, but when electricity flows through it, the gas glows a bright red color.

Argon

Symbol: Ar • Atomic number: 18

Below helium on the periodic table is argon. Argon comprises about 0.02 percent of the universe but is, oddly enough, the third most common gas found in Earth's atmosphere. Because argon is so plentiful on Earth, it is

How Are the Elements Named?

Discovering a new element is not something that happens every day. When someone does find a new element, it's a big deal. Sir William Ramsay discovered four of the noble gases and helped isolate others that were already known. That is a lot of work for one person. If you are lucky enough to discover an element, you are allowed to name it. The same holds true for many discoveries, such as planets, stars, and even diseases. When scientists discover a new element, they often use Greek words as the basis for the name. This is not always the case, however. The element einsteinium is named after the famous scientist Albert Einstein. Also, californium is named after the state of California. The decision about what to name an element is up to the person or people who discovered it.

the most inexpensive of the noble gases to obtain and use. Argon gas can be found in the air around you. To get pure argon out of the atmosphere, air must be chilled and pressurized into a liquid. When this liquid is slowly allowed to warm up, nitrogen will boil away first. As the temperature gets warmer, a mixture of mostly argon with a little oxygen is produced. The oxygen can be removed by passing the mixture over hot copper (Cu), and the oxygen reacts with the copper. Eventually only pure argon will remain.

Krypton

Symbol: Kr • Atomic number: 36

Below argon and above xenon is the noble gas krypton. You may have already heard the name krypton from the comics. It is perhaps best

In the film *Superman Returns*, Lex Luthor uses kryptonite to try to defeat Superman. This mythical radioactive crystal is a fragment from Superman's destroyed home world, Krypton.

known not as an element or a noble gas but as the name of Superman's home planet. In the comic, krypton is also the source of Superman's one weakness, the green radioactive crystals known as kryptonite. But that is all fiction. In reality, krypton is one of the noble gases and makes up only 0.0001 percent of the air on Earth. It has many uses, but its practical value is somewhat lessened due to its rarity, which makes it very expensive. In its solid form, krypton is white and crystal-like. It can be seen in the light spectrum as a brilliant green and orange color.

Xenon

Symbol: Xe • Atomic number: 54

Below krypton on the periodic table is xenon. Xenon is extremely rare on Earth. For every billion atoms in Earth's atmosphere, only about ninety are xenon atoms. The gas is much more common on the sun and on Jupiter. It is found on Mars in about the same amount that it is found on Earth. Xenon gives off a blue or lavender glow when electricity passes through it. Despite its scarcity, xenon is extracted from the atmosphere and is used commercially. Due to its rarity in Earth's atmosphere, xenon was the last noble gas to be discovered. Scottish chemist Sir William Ramsay and English chemist Morris W. Travers discovered it in 1898.

Radon

Symbol: Rn • Atomic number: 86

The final noble gas is radon. It sits at the bottom of group 0 on the periodic table. Radon is a rare, radioactive gas. When an element is radioactive, it means the atoms making it up are unstable and give

A radon test can be used at home and then sent to a laboratory for analysis. It detects the presence of the radioactive and harmful noble gas radon.

off energy particles as they break apart or decay. When radon is cooled enough, below -160°F (-71°C), it solidifies and begins to glow yellow. It changes to red-orange at even lower temperatures. Radon is the only radioactive gas and can be found underground near rock containing the radioactive element radium, which decays into radon over time. Radon escapes from the ground and can sometimes build up in the still air of basements in homes and buildings. It is considered a toxic hazard if found in high quantities over time. Testing for the presence of radon gas before buying a new home has become common in the United States. Eventually, radon will decay into the element polonium (Po).

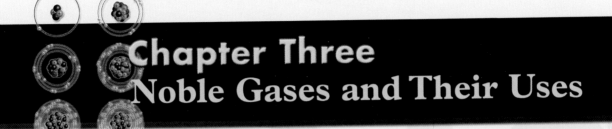

Chapter Three
Noble Gases and Their Uses

Since noble gases are so nonreactive with other elements, and even among themselves, you might think there would be little practical use for such elements. In fact, it is the very nonreactive nature of the noble gases that often makes them very useful. The noble gases may be rare in the atmosphere and hard to isolate, but you might be surprised at all the everyday uses they have. In fact, it is very likely that there are noble gases in your home right now and you're completely unaware it.

Helium and Its Uses

Helium is best known as the gas that makes balloons float. It is the lightest of all the noble elements. It is even lighter than air. So when balloons are filled with helium, they will rise upward in the air, just like air bubbles would in water. Helium is also used in blimps (also known as airships), hot air balloons, and weather balloons to make them rise and float in the air.

In the past, hydrogen was used in airships because it is even lighter than helium and can carry much more weight. Hydrogen is no longer used for airships, however, because it is very explosive and flammable. Today, only helium is used for blimps. Helium-filled advertising blimps can often be seen circling the skies at sporting events.

Helium-filled hot air balloons can transport people short distances. Some balloons even take part in races. This race took place in Albuquerque, New Mexico.

Helium is also used in gas lasers, which produce a bright red beam of light when electricity passes through the gas. The light from helium is passed back and forth between mirrors, producing a very narrow beam of light from a specific color.

Liquid helium gets extremely cold but does not become a solid. It is used to cool the large magnets in particle accelerators. Rockets such as those used by NASA also use helium to help keep the other flammable gases cold. Helium is a good material for these uses because it will not react with the explosive rocket fuel.

You may be aware that helium, when inhaled, will also change the sound of your voice for a short period of time. Since it is less dense than air, helium will raise the pitch of your voice if inhaled from a party balloon. Your voice will become high pitched and squeaky sounding until the helium exits your mouth as you speak. It is not recommended to do this, however. Because inhaling helium can reduce the amount of oxygen in the lungs, it can be dangerous if done to excess.

Neon and Its Uses

Neon is probably the most famous of all the noble gases. Most of us have seen neon lights used in colorful storefront windows. Carnivals, amusement parks, theaters, and restaurants commonly use this gas in tubes

Without the noble gas neon, carnivals and other nighttime amusements would be far less colorful and visually spectacular.

with electricity passing through them. When an electrical current is applied to low pressure neon in a closed tube, the neon electrons become excited and radiate a bright red-orange glow. Although the name "neon" is used to describe the variety of colorful tube lights seen at night, they do not all use neon gas. Some use argon with mercury vapor, which produces blue light. Other colors are produced by coating the inside of the tube with a material that converts the light from mercury to light of a different color.

Helium vs. Hydrogen: The *Hindenburg* Disaster

The *Hindenburg* was a luxury German airship that flew in the mid-1930s. It was able to fly because the large balloon that carried it upward was filled with buoyant hydrogen. However, pure hydrogen gas is extremely flammable. On May 6, 1937, the famous airship caught fire, crashed, and burned while trying to dock in Lakehurst, New Jersey. Thirty-five of its ninety-seven passengers were killed, as was a member of the ground crew. Experts agree that if helium had been used instead of hydrogen, the airship would not have been as flammable and may not have exploded. In the 1930s, helium was available but only from the United States. When the *Hindenburg* flew, the United States would not allow sales of helium to Germany because of that nation's military aggressions in pre–World War II Europe. The Germans, therefore, used hydrogen instead of helium. Today, blimps and airships, which are rarely used as passenger vehicles anymore, no longer use hydrogen. The noble, and far safer, gas helium is used instead.

Argon and Its Uses

Argon is most commonly used in conventional incandescent lightbulbs. The bulb's thin metal wires become hot and glow when electricity is passed through them. In normal oxygen-rich air, the wires would burn up. However, modern lightbulbs are filled with argon gas, which is unreactive, so the hot wire can glow brightly without burning.

Argon is very abundant in our atmosphere. It is also very inexpensive to produce for industrial uses. Another popular use for this gas is in arc welding. In welding, two or more metal pieces are joined by applying

As lightbulbs are manufactured, a mixture of argon and nitrogen is used to replace the air inside the bulb so that the filament will not burn out as quickly.

extreme heat. As a welder is working, the inert gas argon is continually blown over the hot weld so that air cannot react with the hot, molten metal being produced. This prevents the metals from burning or causing a fire.

Argon is much heavier than air and is often injected between sealed panes of glass and used as insulation in modern windows. This insulation makes the windows very energy efficient, preventing heat from either escaping (in winter) or entering (in summer) the house.

An atmosphere of pure argon can help produce certain industrial materials. Silicon (Si) crystals for the semiconductor industry are grown inside chambers filled with inert argon. Titanium (Ti) production also requires an inert argon atmosphere. Argon lasers, which produce green light, have a variety of uses in the medical industry. For example, some of them are used for the cosmetic removal of birthmarks, as well as in corrective eye surgery.

Krypton and Its Uses

Like argon, krypton is also used in incandescent lightbulbs. However, krypton is more expensive to produce than argon, so it is not used as often. It is present, along with argon, in fluorescent lights. Krypton lights glow very brightly because they convert electrical energy to light energy quite well. Because of this brightness and high efficiency, they are used most often in airport runway lights. Pilots rely on these bright lights to help them guide their planes while landing in the dark. Krypton is also mixed with xenon for use in strobe lights and flashbulbs for photography.

Xenon and Its Uses

Like other noble gases, xenon emits light when electricity is passed through it. Xenon is used and mixed with other gases in lights and lighting equipment. Photoflashes light up a scene and allow a camera to

Xenon lightbulbs are used in lighthouses because xenon gas glows extremely brightly when electricity is passed through it.

capture and "freeze" a fraction of a second in time, such as the splash of a water droplet on the ground. Strobe lights and camera flashbulbs contain a mixture of xenon and krypton. In addition, xenon lamps are so bright that they are used in lighthouses to help guide ships sailing near the shore.

Radon and Its Uses

Radon is very radioactive. As a result, it has very few practical uses. For many years, doctors have used radon to kill cancer cells in humans, in a process known as radiation. Small amounts of radon are put into small glass or gold capsules, called seeds, and implanted in the patient at the area of the cancer growth. Although radon is used to treat cancer patients, exposure to radon and radiation is not safe for humans. It kills cancer cells, but it also kills healthy cells and makes the patient ill. Because radon can naturally occur in the ground, homeowners often place radon detectors in their basement to identify any radon gas seeping up from the ground into their house. Although radon can be used in the fight against cancer, long-term uncontrolled exposure to the gas can actually cause cancer. In fact, exposure to radon is the second-leading cause of lung cancer after smoking.

How Are the Noble Gases Isolated?

The air around you contains many gases. It is made up of nitrogen, oxygen, carbon dioxide, and several of the six noble gases. The air that we breathe is a mixture, not a compound. However, separating these gases from one another so that they can be used for our purposes can be quite complicated.

A process called fractional distillation is used to separate the gases in the air from one another. First, the air is compressed and

This fractional distillation tower is used for refining oil into gasoline. Similar distillation towers are used for refining various gases out of the air we breathe.

put under a lot of pressure. Then, the air is cooled so that water vapor and carbon dioxide will freeze into ice crystals that can be removed. Then, the air is cooled even more. Eventually, most of the gases still present become liquid at about -392°F (-200°C).

The pressurized liquid mixture is put into the top of a large tower. It is here that the remaining gases are separated from one another. The tower is colder at the top and warmer at the bottom, and it has trays with holes in them that the liquid air runs through. As the cold liquid air trickles through the holes, it gradually becomes warmer. Eventually it gets warm enough that the nitrogen in the liquid turns back into gas. The nitrogen gas can be collected near the top of the tower. Farther down, the liquid gets warm enough that the argon turns from liquid to gas. Argon gas is removed near the middle of the tower. Near the bottom of the tower, the liquid gets warm enough that oxygen turns to gas, and it is removed. Each gas is collected and placed in its own separate tank for further purifying.

Most noble gases can't be isolated and collected using this fractional distillation process. The first gas to form when liquid air is warmed is a mixture of hydrogen, neon, and helium. Hydrogen is very flammable, so it is burned to remove it from the mixture. The helium is removed from the neon with a special form of carbon (C) called activated carbon. Activated carbon is a very porous form of charcoal that can absorb gas. Helium gas is not usually produced for industry through this process. In the United States, several underground caverns where natural gas is present also contain helium. The helium is extracted using activated carbon to remove the other gases. Radon, the radioactive noble gas, is collected during the radioactive decay of minerals that contain radium.

Chapter Four
Compounds and the Noble Gases

The noble gases get their name because they are known for being very "antisocial" elements, keeping to themselves and not readily reacting with the other elements around them. Originally, scientists thought it was impossible to make any compounds from these six elements: helium, neon, argon, krypton, xenon, and radon. Their nonreactive nature and complete outer electron shells seemed to make it impossible for any chemical reaction to occur that would allow a compound to be created.

However, almost as soon as the noble elements were discovered and thought to be entirely nonreactive, scientists got to work trying to find the cracks in this theory by making a compound in the laboratory. Heavier elements have a weaker attraction between the nucleus and the outer shell of their electrons, increasing the likelihood of a chemical reaction. Therefore, the heaviest of the noble gases, radon and xenon, showed the greatest promise for forming compounds, even though their outer shells are complete. Scientists hoped they could grab an electron from the outer shell of a radon or xenon atom and create a situation in which the electron could become involved in forming a bond with an atom of another element. Scientists worked for many years trying to create this kind of a compound and break the heavier noble elements out of their apparent isolation.

A photon source plasma bulb like the one shown here often contains helium and neon gases. When electricity flows through the sphere, the gases glow. The gases react in a variety of ways when the outside of the bulb is touched.

How do scientists get atoms with complete outer shells to give up an electron? They have experimented with many methods. They can excite the atoms with great amounts of electricity. They can cause atoms to collide at high speeds. These conditions can only be created in a laboratory under highly controlled conditions. That is what makes experimentation difficult and often very expensive.

First Successes

In 1962, the chemist Neil Bartlett finally did what was once believed to be impossible. While experimenting with a compound called platinum

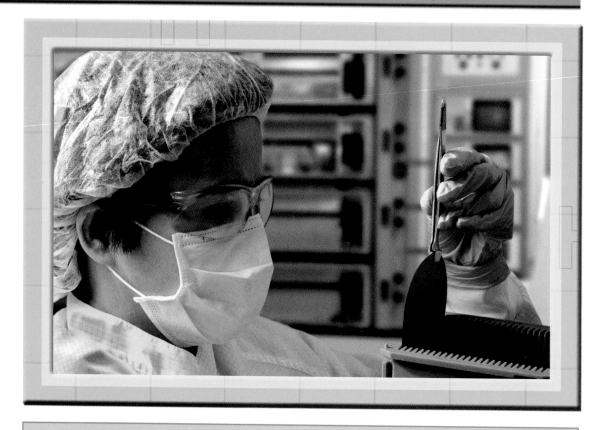

This scientist is inspecting a silicon wafer, used in computer technology. The wafers are grown in inert noble gas chambers.

hexafluoride, Bartlett became the first to successfully create a compound with the noble gas xenon. The resulting compound is xenon platinum hexafluoride, which is a red, solid, crystal-like compound. It has no commercial or practical use but was the first breakthrough in forming compounds with noble gases. It was proof that the noble elements could in fact be chemically combined with other elements.

Since then, many xenon compounds have been created. One of the most stable and useful is xenon difluoride. This compound is used to etch silicon, which is helpful in the semiconductor industry in the production of microelectromechanical systems (MEMS). MEMS are very small, even microscopic machines.

Mixture vs. Compound

We know that noble gases do not easily combine with other elements to form compounds. But in a mixture with other elements, they can be used for many purposes. What is the difference between a mixture and a compound? A compound is a chemical combination of elements that is made during a reaction between the elements. Each of the elements combine on a molecular level, and the resulting compound has new properties. A mixture is a physical combination of elements in which each element keeps its own properties. Air is a mixture of several gas elements, including oxygen, while water is a compound of hydrogen and oxygen. You can breathe the oxygen in air; you can't breathe the oxygen in water.

As a result of Bartlett's work, more xenon compounds have been made. Some have very specific uses in industry. But most compounds created with noble gases are extremely unstable. That means they decompose when they come in contact with many other materials. Most will explode in air. Because of this they are found only in laboratories where they can be handled properly.

Breaking Up Atoms

At first, attempts to create compounds from noble gases received little attention and funding. People thought that, similar to the xenon platinum hexafluoride that Bartlett discovered, there would be no practical use for any compounds that could be created. Today, however, more attention is being given to noble gas compounds, and interesting uses are being explored for some of them. For example, the elements chlorine (Cl) and fluorine (F) have been chemically combined in laboratories to react with argon, krypton, and xenon. In order for a compound to be created from this combination, however, a powerful electrical current must be used. The result is a very unstable molecule that exists for less than a second before breaking apart again. Can there be a practical use for such a short-lived molecule?

Scientists think there can be. These temporary new molecules, called excimers, release pulses of powerful energy when they break apart. These potent bursts of energy are being explored for use in lasers, weapons, and missile shields. It is the very unstable nature of excimers that make them so interesting to scientists. Indeed, their usefulness resides in their weakness as molecules. They are most powerful at the moment they fall apart.

Hope for the Future

Scientists still have a long way to go in the creation of more compounds from noble gases. As they continue to try to form new chemical bonds,

Neil Bartlett

The chemist Dr. Neil Bartlett was born in England in 1932. He made his famous discovery of the first noble gas compound while working at the University of British Columbia in Vancouver, Canada. He later moved to the United States, where he conducted research at Princeton University and the University of California, Berkeley. In addition to this discovery, Bartlett was famous among chemists for his many experiments with the element fluorine. He died in 2008.

Photographers use flashbulbs filled with krypton and xenon gases.

they are also looking for ways to make the existing ones more stable. If chemists can create compounds that stay together long enough to be studied, they may find practical uses for them. Often when scientists discover something like a new compound, it does not always have an immediately obvious use. Sometimes it takes years, decades, or even centuries for practical uses to be found for scientific discoveries. In fact, more often than not no practical uses are found for scientific discoveries.

The noble gases continue to hold great wonder and extreme challenge for scientists. One day, science may find a way to create and use many noble gas compounds. For now, however, we simply appreciate the everyday uses of these noble elements in their pure form, in lightbulbs, balloons, and neon signs, and in such industries as welding and photography.

The Periodic Table of Elements

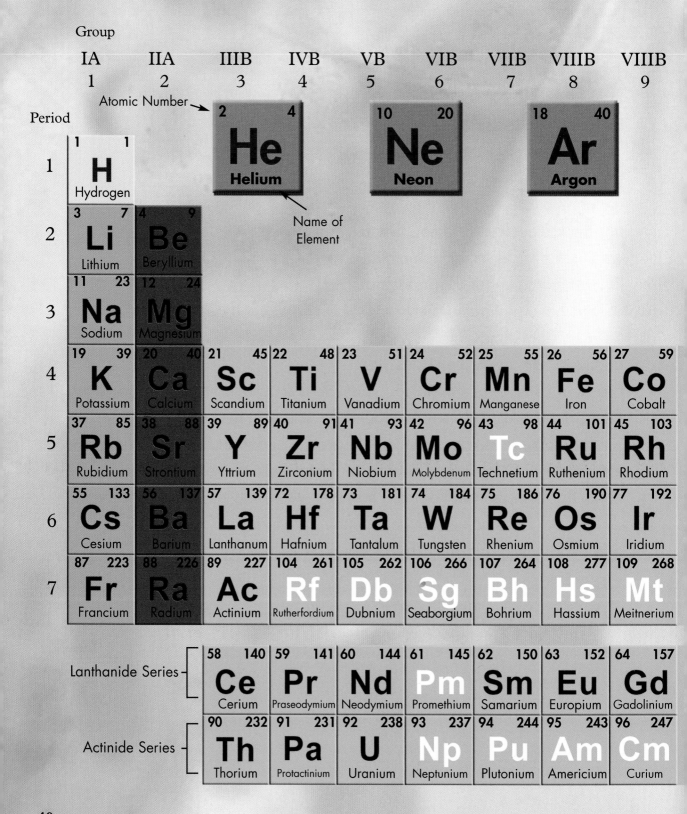

Group

	IA	IIA	IIIB	IVB	VB	VIB	VIIB	VIIIB	VIIIB
	1	2	3	4	5	6	7	8	9

Atomic Number →

2	4
He	
Helium	

10	20
Ne	
Neon	

18	40
Ar	
Argon	

Name of Element

Period

1
1	1
H	
Hydrogen	

2
3	7	4	9
Li		**Be**	
Lithium		Beryllium	

3
11	23	12	24
Na		**Mg**	
Sodium		Magnesium	

4
19	39	20	40	21	45	22	48	23	51	24	52	25	55	26	56	27	59
K		**Ca**		**Sc**		**Ti**		**V**		**Cr**		**Mn**		**Fe**		**Co**	
Potassium		Calcium		Scandium		Titanium		Vanadium		Chromium		Manganese		Iron		Cobalt	

5
37	85	38	88	39	89	40	91	41	93	42	96	43	98	44	101	45	103
Rb		**Sr**		**Y**		**Zr**		**Nb**		**Mo**		**Tc**		**Ru**		**Rh**	
Rubidium		Strontium		Yttrium		Zirconium		Niobium		Molybdenum		Technetium		Ruthenium		Rhodium	

6
55	133	56	137	57	139	72	178	73	181	74	184	75	186	76	190	77	192
Cs		**Ba**		**La**		**Hf**		**Ta**		**W**		**Re**		**Os**		**Ir**	
Cesium		Barium		Lanthanum		Hafnium		Tantalum		Tungsten		Rhenium		Osmium		Iridium	

7
87	223	88	226	89	227	104	261	105	262	106	266	107	264	108	277	109	268
Fr		**Ra**		**Ac**		**Rf**		**Db**		**Sg**		**Bh**		**Hs**		**Mt**	
Francium		Radium		Actinium		Rutherfordium		Dubnium		Seaborgium		Bohrium		Hassium		Meitnerium	

Lanthanide Series
58	140	59	141	60	144	61	145	62	150	63	152	64	157
Ce		**Pr**		**Nd**		**Pm**		**Sm**		**Eu**		**Gd**	
Cerium		Praseodymium		Neodymium		Promethium		Samarium		Europium		Gadolinium	

Actinide Series
90	232	91	231	92	238	93	237	94	244	95	243	96	247
Th		**Pa**		**U**		**Np**		**Pu**		**Am**		**Cm**	
Thorium		Protactinium		Uranium		Neptunium		Plutonium		Americium		Curium	

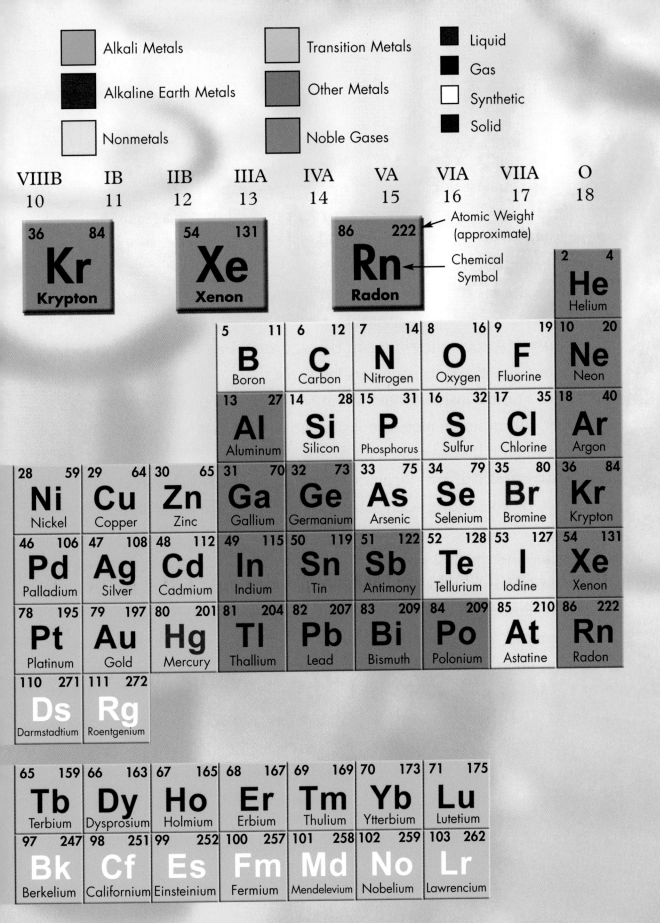

Alkali Metals
Alkaline Earth Metals
Nonmetals
Transition Metals
Other Metals
Noble Gases
Liquid
Gas
Synthetic
Solid

VIIIB 10	IB 11	IIB 12	IIIA 13	IVA 14	VA 15	VIA 16	VIIA 17	O 18

Atomic Weight (approximate)
Chemical Symbol

| 36 84 **Kr** Krypton | | 54 131 **Xe** Xenon | | 86 222 **Rn** Radon | | | | |

| | | | 5 11 **B** Boron | 6 12 **C** Carbon | 7 14 **N** Nitrogen | 8 16 **O** Oxygen | 9 19 **F** Fluorine | 2 4 **He** Helium |

| | | | 13 27 **Al** Aluminum | 14 28 **Si** Silicon | 15 31 **P** Phosphorus | 16 32 **S** Sulfur | 17 35 **Cl** Chlorine | 10 20 **Ne** Neon |

| 28 59 **Ni** Nickel | 29 64 **Cu** Copper | 30 65 **Zn** Zinc | 31 70 **Ga** Gallium | 32 73 **Ge** Germanium | 33 75 **As** Arsenic | 34 79 **Se** Selenium | 35 80 **Br** Bromine | 18 40 **Ar** Argon |

| 46 106 **Pd** Palladium | 47 108 **Ag** Silver | 48 112 **Cd** Cadmium | 49 115 **In** Indium | 50 119 **Sn** Tin | 51 122 **Sb** Antimony | 52 128 **Te** Tellurium | 53 127 **I** Iodine | 36 84 **Kr** Krypton |

| 78 195 **Pt** Platinum | 79 197 **Au** Gold | 80 201 **Hg** Mercury | 81 204 **Tl** Thallium | 82 207 **Pb** Lead | 83 209 **Bi** Bismuth | 84 209 **Po** Polonium | 85 210 **At** Astatine | 54 131 **Xe** Xenon |

| 110 271 **Ds** Darmstadtium | 111 272 **Rg** Roentgenium | | | | | | | 86 222 **Rn** Radon |

65 159 **Tb** Terbium	66 163 **Dy** Dysprosium	67 165 **Ho** Holmium	68 167 **Er** Erbium	69 169 **Tm** Thulium	70 173 **Yb** Ytterbium	71 175 **Lu** Lutetium
97 247 **Bk** Berkelium	98 251 **Cf** Californium	99 252 **Es** Einsteinium	100 257 **Fm** Fermium	101 258 **Md** Mendelevium	102 259 **No** Nobelium	103 262 **Lr** Lawrencium

Glossary

atom The smallest stable form of matter that exists.

atomic mass The mass of a single atom and all its parts.

atomic number The number that identifies an atom and represents the number of protons in an atom.

chemical element A substance that cannot be separated into other substances.

compound A substance that can be separated into elements, and the ratio of those elements is always the same.

electron Particle inside an atom that has a negative electrical charge.

excimer An unstable molecule that releases strong electromagnetic pulses when it breaks down.

fractional distillation A process used to separate substances in a mixture by vaporizing each substance at a different temperature.

inert Describing something that is unmoving.

molecule The smallest particle of a substance, usually containing two or more atoms bonded to each other.

neutron Particle inside an atom that has no electrical charge.

noble gases A group of elements on the periodic chart that includes helium, neon, argon, krypton, xenon, and radon. They do not easily combine with other elements to form compounds.

nucleus The center part, or core, of an atom that contains protons and usually neutrons, and that contains nearly all the mass of the atom.

proton Particle inside an atom that has a positive electrical charge.

viscosity The measure of internal friction of liquids or fluids.

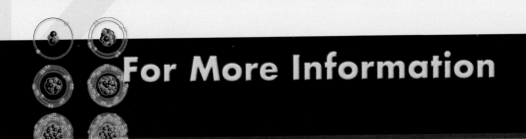

American Chemical Society (ACS)
1155 Sixteenth Street NW
Washington, DC 20036
(800) 227-5558
Web site: http://www.acs.org
The ACS supports professionals in all areas of chemistry and students
   interested in science.

Canadian Society for Chemistry (CSC)
130 Slater Street, Suite 550
Ottawa, ON K1P 6E2
Canada
(613) 232-6252
Web site: http://www.cheminst.ca
This national technical association represents the field of chemistry and
   the interests of chemists in industry, academia, and government.

Knowledge Learning Corporation School Partnerships
10106 West San Juan Way, Suite 100
Littleton, CO 80127
(888) 458-1812
Web site: http://www.klcschoolpartnerships.com
The KLC School Partnerships provide school and community services
   that include the Science Adventures summer day camps.

National Science Foundation (NSF)
4201 Wilson Boulevard

Arlington, VA 22230
(703) 292-5111
Web site: http://www.nsf.gov
The NSF is a federal agency that supports many science education programs for students.

National Youth Science Organization
P.O. Box 3387
Charleston, WV 25333-3387
(304) 342-3326
Web site: http://www.nysc.org
This nonprofit organization supports and operates science education
programs, including the National Youth Science Camp in West Virginia.

The Science Club
4921 Preston/Fall-City Road
Fall City, WA 98024
(425) 222-5066
Web site: http://www.scienceclub.org
This nonprofit organization organizes school assemblies, workshops,
and programs for television, museums, and print sources.

Web Sites

Due to the changing nature of Internet links, Rosen Publishing has
developed an online list of Web sites related to the subject of this book.
This site is updated regularly. Please use this link to access this list:

http://www.rosenlinks.com/uept/noble

For Further Reading

Dingle, Adrian. *The Periodic Table: Elements with Style*. Boston, MA: Kingfisher, 2007.

Newmark, Ann. *Chemistry* (DK Eyewitness Books). New York, NY: DK Children, 2005.

Saunders, Nigel. *Neon and the Noble Gases*. Oxford, England: Heinemann Library, 2003.

Slade, Suzanne. *Elements and the Periodic Table* (Library of Physical Science). New York, NY: PowerKids Press, 2007.

Thomas, Jens. *The Elements: Noble Gases*. White Plains, NY: Benchmark Books, 2003.

Tocci, Salvatore. *Hydrogen and the Noble Gases* (A True Book). New York, NY: Children's Press, 2004.

Tocci, Salvatore. *The Periodic Table* (A True Book). New York, NY: Children's Press, 2004.

Wiker, Benjamin. *The Mystery of the Periodic Table* (Living History Library). Bathgate, ND: Bethlehem Books, 2003.

Bibliography

AbsoluteAstronomy.com. "Henry Cavendish." Retrieved January 26, 2009 (http://www.absoluteastronomy.com/topics/Henry_Cavendish).

Asimov, Isaac. *The Noble Gases*. New York, NY: Basic Books, 1966.

Biography Base. "Neal Bartlett Biography." Retrieved January 9, 2009 (http://www.biographybase.com/biography/Bartlett_Neil.html).

Browne, Malcolm W. "On Elements That Want to Be Alone." *New York Times*, March 2, 1988. Retrieved January 7, 2009 (http://query.nytimes.com/gst/fullpage.html?res=940DE1D91730F931A15750C0A96E948260&scp=2&sq=noble%20metals&st=cse).

Dingle, Adrian. *The Periodic Table: Elements with Style*. Boston, MA: Kingfisher, 2007.

Heiserman, David L. *Exploring Chemical Elements and Their Compounds*. New York, NY: TAB Books, 1992.

Kiwi Web. "Who Was Dmitri Mendeleev?" Retrieved December 29, 2008 (http://www.chemistry.co.nz/mendeleev.htm).

Newton, David E. *The Chemical Elements*. New York, NY: Franklin Watts, Inc., 1994.

New York Times Company. *New York Times Guide to Essential Knowledge*. 2nd ed. New York, NY: St. Martin's Press, 2007.

NobelPrize.org. "Sir William Ramsey: The Nobel Prize in Chemistry, 1904." Retrieved January 7, 2009 (http://nobelprize.org/nobel_prizes/chemistry/laureates/1904/ramsay-bio.html).

Saunders, Nigel. *Neon and the Noble Gases*. Oxford, England: Heinemann Library, 2003.

Stwertka, Albert. *A Guide to the Elements*. 2nd ed. New York, NY: Oxford University Press, 2002.

Thomas, Jens. *The Elements: Noble Gases*. White Plains, NY: Benchmark Books, 2003.

World Almanac for Kids. "It All Starts with an Atom." Retrieved December 28, 2008 (http://www.worldalmanacforkids.com/WAKI-ViewArticle.aspx?pin=wak-242005&article_id=525&chapter_id=12&chapter_title=Science&article_title=Atoms).

Index

About the Author

Adam Furgang has written several books for Rosen Publishing. He has always had an interest and appreciation for all things scientific.

Photo Credits

Cover, pp. 1, 8, 11, 40–41 by Tahara Anderson; p. 5 © doc-stock/Corbis; p. 12 © Bettmann/Corbis; p. 13 © Kit Kittle/Corbis; p. 16 © Dr. Tim Evans/Photo Researchers; p. 18 © Richard Treptow/Photo Researchers; p. 19 © Richard Cummins/Corbis; p. 21 © Everett Collection; p. 23 © Charles D. Winters/Photo Researchers; p. 26 © Steve Snowden/Getty Images; p. 27 © Atlantide Phototravel/Corbis; p. 29 © Mark Edwards/Peter Arnold, Inc.; p. 31 © www.istockphoto.com/Bradley Mason; p. 32 © Paul Rapson/Photo Researchers; p. 35 © Johannes Simon/Getty Images; p. 36 © www.istockphoto.com/Eliza Snow; p. 39 © www.istockphoto.com/Brett Hillyard.

Designer: Tahara Anderson; Photo Researcher: Marty Levick